Community Helpers at Work

BOOKWORMS

I0884159

A Day with a Firefighter

By Katie Kawa

Cavendish Square

New York

Published in 2021 by Cavendish Square Publishing, LLC
243 5th Avenue, Suite 136, New York, NY 10016

First Edition

Website: cavendishsq.com

Library of Congress Cataloging-in-Publication Data

Names: Kawa, Katie, author.
Title: A day with a firefighter / Katie Kawa.
Description: New York : Cavendish Square Publishing, [2021] |
Series: Community helpers at work | Includes index.
Identifiers: LCCN 2019045637 (print) | LCCN 2019045638 (ebook) |
ISBN 9781502658142 (paperback) | ISBN 9781502658159 (set) |
ISBN 9781502658166 (library binding) | ISBN 9781502658173 (ebook)
Subjects: LCSH: Fire extinction–Juvenile literature. |
Fire fighters–Juvenile literature.
Classification: LCC TH9148 .K388 2021 (print) | LCC TH9148 (ebook) |
DDC 628.9/2–dc23
LC record available at https://lccn.loc.gov/2019045637
LC ebook record available at https://lccn.loc.gov/2019045638

Editor: Katie Kawa
Copy Editor: Nathan Heidelberger
Designer: Andrea Davison-Bartolotta

The photographs in this book are used by permission and through the courtesy of: Cover Sergey Mironov/Shutterstock.com; p. 5 Toa55/Shutterstock.com; p. 7 Tyler Olson/Shutterstock.com; p. 9 © iStockphoto.com/stevecoleimages; p. 11 D-VISIONS/ Shutterstock.com; p. 13 Geet Theerawat/Shutterstock.com; p. 15 potowizard/Shutterstock.com; p. 17 My Portfolio/Shutterstock.com; p. 19 Monkey Business Images/Shutterstock.com; p. 21 Chris Briggs/Shutterstock.com; p. 23 wavebreakmedia/Shutterstock.com.

CPSIA compliance information: Batch #CS20CSQ: For further information contact Cavendish Square Publishing LLC, New York, New York, at 1-877-980-4450.

Printed in the United States of America

CONTENTS

Fighting Fires

Firefighters work very hard. They help people in their **community** stay safe. They do this by putting out fires. They help people in other ways too. Firefighters are very brave!

Firefighters work at a place called a fire station. Some people also call it a firehouse. Firefighters keep many of their tools and clothes there. Sometimes they sleep at the fire station!

When there's a fire, a firefighter has to act quickly. They put on clothes and boots that keep them safe in a fire. They put on **gloves** and a **helmet** too.

12
FIREFIGHTER

A firefighter often rides in a fire truck to get to a fire. A fire truck can go very fast! It has colored lights on top and a loud **siren**. Most fire trucks are red.

ROUTE
66
8

A firefighter often uses a hose. A hose carries water. A firefighter uses water from the hose to put out a fire. Sometimes a firefighter needs a lot of water to put out a big fire!

A firefighter uses other tools too. They use a ladder to reach high places. This is helpful when there's a fire in a tall building. A firefighter can use the ladder to help people who are stuck.

Firefighters work in many different places. Forest fires can cause big problems. They kill many trees. They can also burn down many homes if they start to **spread**. Firefighters work hard to stop forest fires.

Saving People

Firefighters get people out of buildings or cars that are on fire. Firefighters help people if they're hurt too. Firefighters know first aid. This is a way to care for people who are sick or hurt.

Other Ways to Help

When there are no fires to fight, firefighters still work hard. They check the fire truck and keep it clean. They help people who are sick or hurt in their community. They help animals too.

T25

Firefighters visit schools and other places in their community. They teach people about fire safety. Kids can visit a fire station and meet a firefighter. That's a fun way to learn more about this important job!

WORDS TO KNOW

community: An area where people live; a neighborhood.

gloves: Coverings for the hands with parts for each finger.

helmet: A hard hat that is worn to keep the head safe.

siren: A device that makes a loud warning sound.

spread: To get bigger and move into more places.

INDEX